Reactivity and Structure
Concepts in Organic Chemistry

Volume 18

Editors:

Klaus Hafner   Jean-Marie Lehn
Charles W. Rees   P. von Ragué Schleyer
Barry M. Trost   Rudolf Zahradnik

D. F. Taber

Intramolecular Diels-Alder and Alder Ene Reactions

Springer-Verlag
Berlin Heidelberg New York Tokyo 1984

Douglass F. Taber
Department of Chemistry
University of Delaware
Newark, Delaware 19711/USA

List of Editors

Professor Dr. Klaus Hafner
Institut für Organische Chemie der TH
Petersenstr. 15, D-6100 Darmstadt

Professor Dr. Jean-Marie Lehn
Institut de Chimie, Université de Strasbourg
1, rue Blaise Pascal, B.P. 296/R 8, F-67008 Strasbourg-Cedex

Professor Dr. Charles W. Rees, F. R. S. Hofmann
Professor of Organic Chemistry, Department of Chemistry
Imperial College of Science and Technology
South Kensington, London SW7 2AY, England

Professor Dr. Paul v. Rague Schleyer
Lehrstuhl für Organische Chemie der Universität Erlangen-Nürnberg
Henkestr. 42, D-8520 Erlangen

Professor Barry M. Trost
Department of Chemistry, The University of Wisconsin
1101 University Avenue, Madison, Wisconsin 53706, U.S.A.

Prof. Dr. Rudolf Zahradník
Tschechoslowakische Akademie der Wissenschaften
J.-Heyrovský-Institut für Physikal. Chemie und Elektrochemie
Máchova 7, 121 38 Praha 2, C.S.S.R.

ISBN-13: 978-3-642-69235-2   e-ISBN-13: 978-3-642-69233-8
DOI: 10.1007/978-3-642-69233-8

Library of Congress Cataloging in Publication Data.
Taber, Douglass F., 1948 —. Intramolecular Diels-Alder and Alder ene reactions.
(Reactivity and structure; v. 18). Bibliography: p. Includes indexes. 1. Diels-Alder
reaction. I. Title. II. Series.
QD281.R5T33 1984 547′.2 83-12559
ISBN-13: 978-3-642-69235-2 (U.S.)

© by Springer-Verlag Berlin Heidelberg 1984
Softcover reprint of the hardcover 1st edition 1984

2152/3020-543210

Preface

The Diels-Alder reaction has long been a powerful tool in organic synthesis. In recent years, the Alder ene reaction has also achieved some prominence. From the beginning, it was apparent that the intramolecular variants of these reactions would be feasible. Many such have been reported, but the results are widely scattered in the chemical literature. This volume is an attempt to synthesize results observed to date, and to suggest directions for future development.

One of the limiting factors in the application of the intramolecular Diels-Alder reaction has been the development of methods for the preparation of the requisite trienes. The first chapter of this volume summarizes methods for the preparation of dienes and dienophiles. Examples representative of every general approach to 1,3-dienes and to dienophilic functional group combinations have been included.

There are two questions one might ask in considering the prospective cyclization of a given triene: what are the factors that govern the rate of cyclization? and, for cyclizations that lead to the creation of one or more new chiral centers, what are the factors that govern diastereoselectivity? These questions are addressed in Chapter Two.

The third chapter is devoted to the all-carbon intramolecular Alder ene reaction. The tables in that chapter summarize all examples that could be found in the literature through 1981, with several additional examples from 1982. Leading references to heterocyclic ene reactions are also included in this chapter.

This volume will, in many senses, complement an Organic Reactions chapter on the intramolecular Diels-Alder reaction recently prepared by Dr. Engelbert Ciganek of the E. I. DuPont Experimental Station. The tables in that chapter summarize all literature examples of the intramolecular Diels-Alder reaction. I am most grateful to Dr. Ciganek for sharing his references with me.

In conclusion, I would like to express my appreciation to Professor Gilbert Stork, from whom I first learned the detailed analysis of diastereomeric transition states that has formed the basis of our work on the intramolecular Diels-Alder reaction, and to Professor Barry Trost, whose imagination and energy inspired this volume. I would also like to thank Ms. Vicki Long, who typed this manuscript, and Ms. Judy Joos, who drew the several hundred illustrations. Finally, I owe the greatest debt to the many researchers whose detailed observations made this volume possible.

Newark, Delaware   Douglass F. Taber
September, 1983

Table of Contents

Chapter 1.

The Intramolecular Diels-Alder Reaction: Variations and Scope

I. Introduction

The Diels-Alder reaction has long been a useful tool in organic synthesis. Although the concept of covalently linking the diene and the dienophile was originated by Alder (1), it has only been in the last ten years that the intramolecular Diels-Alder reaction has been widely used. Partly this has been because efficient routes to the requisite trienes have only recently been available, and partly it has been because there has not been enough data in the literature to allow one to predict with confidence the stereochemical outcome of a prospective cyclization.

The purpose of this chapter is to examine the scope of the intramolecular Diels-Alder reaction, with particular emphasis on methods for assembling the starting triene. While an exhaustive compilation has not been intended, the attempt has been made to include examples of every sort of diene and dienophile, including those with heteroatom substitution.

In the course of a cyclization, up to four new chiral centers are formed. For the intramolecular Diels-Alder reaction to be synthetically useful, one would like to have sufficient understanding of the molecular conformation leading to cyclization that one could predict with confidence the stereochemical outcome of a given example. This topic is addressed in detail in the next chapter.

The scope and mechanism of the intramolecular Diels-Alder reaction have been reviewed (2). While, in general, it is taken to be a concerted [4+2] process, in at least one instance there is substantial evidence for a diradical intermediate [1] (3). Further, even though the cyclization is concerted, it does not follow that the two new bonds are formed simultaneously. Nonsynchronous bond formation has been invoked to explain the stereochemical outcome of some cyclizations (4).

There are six possible cyclization modes for the intramolecular Diels-Alder reaction [2]. When the bridge between the diene and the dienophile is four atoms or less, Mode II is prohibitively strained, and an E triene will cyclize through Mode I. With longer bridges, Mode II becomes available (5). With a Z diene, Modes III and IV are comparable. The proportion of each is then dependent on a variety of factors. These will be discussed in detail in the next chapter. While Modes V and VI, leading to bridgehead olefins, might appear unlikely, preliminary investigations have shown V to be feasible [3]. With a longer bridge, VI might be possible also.

I

II

III

IV

[2]

V

VI

(6) [3]

In addition to these, more complex systems in which the diene and the dienophile are connected by more than one bridge have been investigated [4]. Such systems are of limited preparative flexibility, and so will not be further addressed in these chapters.

(7) [4]

Although it is often the case that the intramolecular Diels-Alder reaction is irreversible, and that the products are therefore kinetically controlled, there are combinations of diene and dienophile for which such is not the case. The balance between starting material and product can then be subtle [5].

(8) [5]

Some of the most elegant applications of the intramolecular Diels-Alder reaction have been in the stereocontrolled construction of complex natural products. Illustrations of such applicability are included throughout these two chapters.

II. Range and Preparation of Dienes

The simplest dienes used in the intramolecular Diels-Alder reaction are acyclic. As with the intermolecular cyclization, E dienes are usually more reactive than Z dienes [6], although with highly activated dienophiles or under forcing conditions the latter can be made to participate in the reaction. This topic is addressed in detail in the next chapter.

(9) [6]

Many diene-containing synthetic fragments are known [7-13]. Employment of such an off-the-shelf moiety can abbreviate the number of steps required for the preparation of a given triene.

(10) [7]

(11) [8]

3

$$(12) \ [9]$$

$$(13) \ [10]$$

$$(14) \ [11]$$

$$(15) \ [12]$$

$$(16) \ [13]$$

The need for differentially-substituted dienes has led to the development of a number of methods for diene construction [14–21]. A critical consideration in the development or application of such a method is the need to control olefin geometry.

$$(17) \ [14]$$

RMgBr + (epoxide) $\longrightarrow$ R~~~OH $\xrightarrow[\text{2)Ph}_3\text{P=CH}_2]{\text{1)PCC}}$ R~~~ (18) [15]

EtO_2P~~~CO_2Me + RCH=O $\longrightarrow$ R~~~CO_2Me (19) [16]

R—C(O)—CH_2—CO_2H + H—C(O)—~~~CO_2Et $\longrightarrow$ R—C(O)~~~CO_2Et (20) [17]

THPO~~~CO_2Me $\xrightarrow{(\quad)_2 CuLi}$ THPO~~~CO_2Me (21) [18]

~~~N + (diester E, E) $\longrightarrow$ (cyclobutene CO_2Et, CO_2Et) $\longrightarrow\longrightarrow$ (22) [19]

(cyclobutene R, dioxolane) $\longrightarrow$ R~~~ (dioxolane) (23) [20]

RCH=O $\xrightarrow{PPh_3}$ R~~~ (23) [20]

$$(24, 25) \ [21]$$

The preceeding examples have all been monosubstituted at the dienophile end of the diene. An internally disubstituted diene has been observed to take part in the intramolecular Diels-Alder reaction, although only after isomerization to the E isomer [22].

$$(26) \ [22]$$

Some dienes have one olefin endocyclic to a ring. One such diene has been prepared by addition of propenylmagnesium bromide to a ketone, followed by dehydration [23]. In another case, such a diene was prepared by an intramolecular aldol condensation, followed by Wittig olefination [24].

$$(27) \ [23]$$

$$(28) \ [24]$$

Finally, a note should be made of the work of Shea, who has investigated Type V cyclizations, leading to bridgehead olefins. The requisite dienes are available from the allylic bromide [25].

(6) [25]

Heterosubstituted acyclic dienes have also been used in the intramolecular Diels-Alder [26–28]. Such dienes can be used to install useful functionality in the newly-formed cyclohexene ring.

(29) [26]

(30) [27]

(31) [28]

7

There are many other routes to acyclic dienes in the chemical literature [29–36]. These may prove useful for the preparation of more complex triene systems.

$$R - \equiv - CO_2R' \longrightarrow R\underset{Me_3Sn}{\overset{CO_2R'}{\diagup}} \longrightarrow \longrightarrow R\underset{Me_3Sn}{\diagup} \qquad (32)\ [29]$$

$$\xrightarrow{Cp_2\,TiCH_2\,AlCl\,Me_2} \qquad (33)\ [30]$$

(reactant: crotonate OEt ester with C=O; product: diene OEt)

$$RCH = O \xrightarrow[\text{TiCl}_4]{H - \equiv -\diagup SiMe_3} R\diagup\diagdown^{Cl} \qquad (34)\ [31]$$

$$\xrightarrow[\text{2) RX}]{\text{1) BuLi}} \xrightarrow[\text{2) NaBH}_4]{\text{1) MeI}} \qquad (35)\ [32]$$

(methylpyridine → 2-(CH$_2$R)pyridine → ...)

$$\xrightarrow[\text{2) KOH}]{\text{1) MeI}} \xrightarrow{Me_2CuLi}$$

(tetrahydropyridine with R → N-diene-R → diene-R)

$$\xrightarrow[\text{2) MeI}]{\text{1) BuLi}}$$

(bicyclic dioxolane–SO$_2$Ph → methylated SO$_2$Ph)

$$\xrightarrow{BuLi} \qquad (36)\ [33]$$

(SO$_2$Ph product structure)

$$RCH = O \ + \ Me_3Si \diagup \diagdown B(\text{pinacol}) \longrightarrow$$

(37) [34]

(38) [35]

$$R_2CuLi \ + \ C_2H_2 \longrightarrow R \diagup \diagdown CuLi \xrightarrow[\text{2)} \ I \diagup \diagdown R']{\text{1) M+}}$$

(39) [36]

Endocyclic dienes ranging in ring size from four to seven have also been used in the intramolecular Diels-Alder reaction. Cyclobutadienes are prepared from the corresponding metal complexes [37]. Cyclopentadienes are most readily prepared by alkylation of cyclopentadienyl anion [38, 39]. They are also available by Michael addition to fulvene derivatives and by aminal exchange [40, 41].

(40) [37]

(41) [38]

9

(42) [39]

(43) [40]

(44) [41]

Cyclohexadienes have most often been derived from cyclohexadienones [42–45]. They are also available by Birch reduction and by Diels-Alder addition followed by elimination [46, 47].

(45) [42]

(46) [43]

(47) [44]

(48) [45]

(49) [46]

11

(50) [47]

1,3-Cycloheptadienes have been prepared from tropyllium derivatives. They are in facile equilibrium, by prototropic shift, one with another, leading in many instances to mixtures of cyclized products [48, 49].

(51) [48]

(52) [49]

18%   58%

12

Heterosubstituted endocyçlic dienes have also been used. In addition to furans [50–53], alkoxydienes [54, 55], benzopyrroles [56], and dihydropyridines [57] have been employed. It should be noted that addition to furans is readily reversible [53]. This difficulty has been overcome by using a pyrone dienophile [54]. Loss of carbon dioxide from the cycloadduct makes the addition irreversible.

(53) [50]

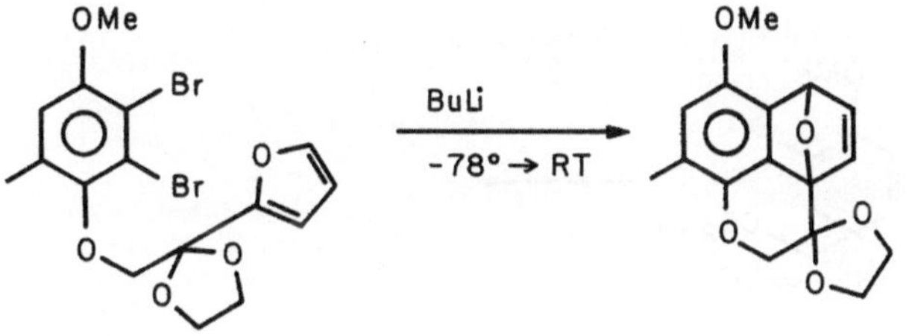

(54) [51]

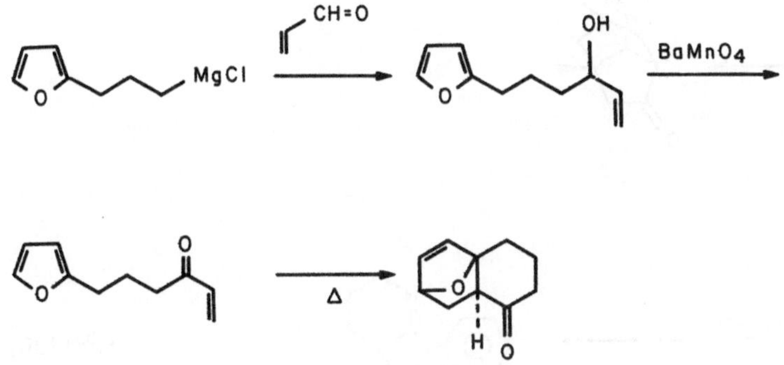

(55) [52]

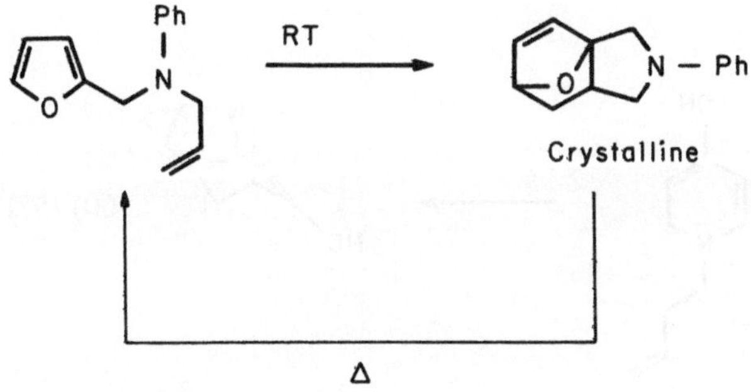

Crystalline

(56) [53]

13

(57) [54]

(58) [55]

(59) [56]

(60) [57]

(61) [58]

With the exception of the abundant o-quinone methide literature (62), there is only a single example of a diene exocyclic to a ring [59]. The use of the very reactive o-quinone methide diene, for intramolecular cycloaddition, was introduced by Oppolzer in 1971 (64). Initially, these were prepared by thermolysis of a preformed benzocyclobutene [60, 61]. Later, Vollhardt demonstrated that 1,5 diacetylenes could serve as precursors to benzocyclobutenes and thus, to o-quinone methides [62]. Since that time, several other methods for the generation of o-quinone methides have been developed [63–68], some of which allow generation of the o-quinone methide under very mild reaction conditions. These methods also allow the incorporation of more complex functionality in the ring system.

(63) [59]

(64) [60]

(65) [61]

(66) [62]

(67) [63]

(68) [64]

(69) [65]

(70) [66]

(71) [67]

(72) [68]

With activated dienophiles, vinyl aromatics can participate in the intra-molecular Diels-Alder reaction [69–75]. (It should be noted that the base-catalyzed examples [73, 74] may be proceeding by electron transfer.) Combination of this approach with the extrusion of carbon dioxide, mentioned above, allows the preparation of a fused aromatic system [75].

(73) [69]

(74) [70]

(75) [71]

(76) [72]

(77) [73]

(78) [74]

(79) [75]

Finally, arenes themselves can serve as dienes [76–81]. When the system is doubly bridged [81] such participation is particularly facile.

(80) [76]

(81) [77]

(82) [78]

19

OAc

tBuOK

(83) [79]

EtO$_2$P〜CO$_2$Et

EtO$_2$C

(84) [80]

Br$_2$/MeOH

0°

(85) [81]

III. Range and Preparation of Dienophiles

As illustrated above, simple olefins and acetylenes, unactivated by electron withdrawing groups, serve well as dienophiles in the intramolecular Diels-Alder reaction. This is in contrast to the intermolecular case. Heterosubstituted olefins can also serve as dienophiles. In addition to the two examples cited above [58, 70] the preparation and cyclization of enamides should be noted [82].

Δ

(10) [82]

Olefins activated by electron withdrawing groups are avid dienophiles. For the preparation of heterocyclic systems, esters and amide dienophiles can be prepared from the (often commercially available) acids [45, 50]. For the preparation of carbocyclic systems, unsaturated esters and ketones have most often been used. Substituted esters are readily prepared by standard procedures [83–88].

$$RCH = O \quad \xrightarrow{\quad Ph_3P = \overset{CO_2Me}{\diagup} \quad} \quad R \diagdown \diagup CO_2Me \qquad \text{(19) [83]}$$

$$R-\equiv-H \quad \longrightarrow \quad R-\equiv-CO_2Me \quad \longrightarrow \quad R \diagdown CO_2Me \qquad \text{(86) [84]}$$

$$RCH = O \quad \xrightarrow{\quad EtO_2P \overset{O}{\overset{\|}{\diagup}} CO_2Me \quad} \quad R \diagdown \diagup CO_2Me \qquad \text{(86) [85]}$$

$$RCH = O \quad \xrightarrow{\quad \overset{Li}{\diagdown}\diagup OEt \quad} \quad R \diagdown \diagup \overset{O \frown Ph}{\diagdown_O} \quad \xrightarrow{\quad MeO_2P \overset{O}{\overset{\|}{\diagdown}} CO_2Me \quad} \quad R \diagdown \diagup CO_2Me \quad ^{O\frown Ph}$$

(87) [86]

$$RCH = O \quad \longrightarrow \quad R \diagdown \overset{O\frown Ph}{\diagup} \equiv\!\!\backsim CO_2Me \quad \xrightarrow{\quad Me_2CuLi \quad} \quad R \diagdown \overset{O\frown Ph}{\diagup} CO_2Me$$

(87) [87]

$$RCH = O \quad \longrightarrow \quad R \diagdown \diagup \overset{CO_2Et}{\diagup}_{Br} \quad \longrightarrow \quad R \diagdown \diagup \overset{OR'}{\diagup}_{Br}$$

(4a) [88]

$$\xrightarrow[\quad MeOH \quad]{\quad Ni(CO)_4 \quad} \quad R \diagdown \diagup \overset{OR'}{\diagup}_{CO_2Me}$$

21

Enones have most often been prepared by oxidation of vinyl carbinols [89, 90]. They have also been prepared by addition of vinyl lithium to a carboxylic acid, and by aldol condensation [91, 92]. In addition, reagents have been developed that allow direct incorporation of the enones [93, 94]. The latter example is an interesting case of a tandem Claisen/Diels-Alder sequence.

(55) [89]

(88) [90]

(49) [91]

(18) [92]

22

(12) [93]

(89) [94]

A variety of other dienophiles have also been investigated. Unsaturated aldehydes [95], amidines [96], nitriles [97] and o-quinones [98] have been used. By combining electron-withdrawing group activation with heterosubstitution, it is possible to directly assemble highly functionalized structures [99–101].

(4c, 87) [95]

(90) [96]

(63) [97]

(91) [98]

(92) [99]

(93) [100]

(94) [101]

Acetylenic esters appear to be avid dienophiles [102, 103]. Klemm [69] has extensively investigated the addition of acetylenic esters to styrene systems. Allenes [104] and ketenes [105] appear to also be active dienophiles.

(95) [102]

(96) [103]

(97) [104]

(98) [105]

As noted above [51], arynes are effective dienophiles. There is a single report, in a doubly bridged system, of an arene serving as a dienophile [106].

(99) [106]

IV. Heterodienes and Dienophiles

Diene/dienophile combinations incorporating one or more hetero atoms have been studied in some detail. The most extensively investigated example has been the further cyclization of one of the condensation products of citral with olivetol [107]. Other o-vinyl phenols cyclize in similar fashion [108, 109]. Unsaturated esters [110], aldehydes [111, 112] and ketones [113] can also serve as dienes.

(100) [107]

(101) [108]

(102) [109]

(103) [110]

(104) [111]

(105) [112]

(106) [113]

27

Diene-containing heterocycles have also been shown to participate in intramolecular cycloaddition [114–119]. Under some circumstances, initial cycloaddition is followed by elimination [117–119]. One other example appears to belong in this category [120].

(107) [114]

(108) [115]

(109) [116]

(110) [117]

(111) [118]

(112) [119]

(113) [120]

Hetero atom-containing dienophiles also take part in the intramolecular Diels-Alder reaction [121–122]. With o-quinone methides [123], cyclization proceeds at the temperature required to generate the diene. The reaction of a nitrile as a dienophile with a simple acyclic diene has also been reported [124].

Extensive work has been reported by Weinreb [125] and by Keck [126] on the preparation and cyclization of more active heterodienophiles.

(114) [121]

(114) [122]

(114) [123]

(115) [124]

(25) [125]

(21) [126]

V. Catalysis of the Reaction

As with bimolecular cycloaddition, the intramolecular Diels-Alder reaction can be catalyzed by Lewis acid [127–129]. In one case, it was reported that trifluoroacetic acid catalyzed cycloaddition [130].

(116) [127]

(117) [128]

(55) [129]

(27) [130]

Other agents have been used to catalyze the bimolecular cycloaddition [131, 132]. It would be interesting to assess the effect of these agents on trienes that undergo thermal cyclization.

(118) [131]

(119) [132]

VI. Conclusion

Looking to the future, with expanding synthetic methodology, it will be possible to efficiently assemble trienes of greater complexity, leading to more highly functionalized products after cycloaddition. Increasingly, the relationship of chiral centers in the triene to those in the cyclized product will be of importance. The factors that govern that relationship, as well as general factors affecting the rate of intramolecular cyclization, are discussed in the next chapter.

VII. References

1. Alder, K., Schumacher, M.: Fortschr. Chem. Org. Naturst. *10*, 66 (1953)
2. a) Oppolzer, W.: Angew. Chem. Int. Ed. (Eng). *16*, 10 (1977)
   b) Carlson, R. G.: Ann. Rep. Med. Chem. *9*, 270 (1974)
   c) Mehta, G.: J. Chem. Ed. *53*, 551 (1976)
   d) Brieger, G., Bennett, J. N.: Chem. Rev. *63* (1980)
3. Martin, S. F. et al.: J. Am. Chem. Soc. *102*, 5274 (1980)

4. a) Boeckman, R. K., Ko, S. S.: ibid. *102*, 7146 (1980)
   b) White, J. D., Sheldon, B. G.: J. Org. Chem. *46*, 2273 (1981)
   c) Taber, D. F. et al.: Tetrahedron Lett. *22*, 5141 (1981)
5. Corey, E. J., Petrzilka, M.: ibid. 2537 (1975)
6. Shea, K. J. et al.: J. Am. Chem. Soc. *102*, 4544 (1980)
7. Wasserman, H. H., Doumaux, A. R.: ibid. *84*, 4611 (1962)
8. Gschwend, H. W. et al.: J. Org. Chem. *41*, 104 (1976)
9. Taber, D. F., Gunn, B. P.: unpublished observations, Vanderbilt University.
10. Martin, S. F. et al.: J. Am. Chem. Soc. *102*, 3294 (1980)
11. Wilson, S. R., Mao, D. T.: ibid. *100*, 6289 (1978)
12. Oppolzer, W., Snowden, R. L.: Tetrahedron Lett. 4187 (1976)
13. Taber, D. F., Saleh, S. A.: J. Am. Chem. Soc. *102*, 5085 (1980)
14. Bailey, S. J. et al.: J. C. S. Chem. Commun. 474 (1978)
15. Oppolzer, W. et al.: Helv. Chim. Acta *60*, 48 (1977)
16. Weinreb, S. M. et al.: J. Am. Chem. Soc. *101*, 5073 (1979)
17. Babler, J. H., Invergo, B. J.: J. Org. Chem. *44*, 3723 (1979)
18. Gras, J. L.: ibid. *46*, 3738 (1981)
19. Roush, W. R.: J. Am. Chem. Soc. *102*, 1390 (1980)
20. Näf, F. et al.: Helv. Chim. Acta *62*, 114 (1979)
21. Keck, G. E., Nickell, D. G.: J. Am. Chem. Soc. *102*, 3632 (1980)
22. Ichihara, A. et al.: ibid. *102*, 6353 (1980)
23. Taber, D. F., Gunn, B. P.: ibid. *101*, 3992 (1979)
24. McIntosh, J. M., Sieler, R. A.: J. Org. Chem. *43*, 4431 (1978)
25. Khatri, N. A. et al.: J. Am. Chem. Soc. *103*, 6387 (1981)
26. Borch, R. F. et al.: ibid. *99*, 1612 (1977)
27. Stork, G. et al.: ibid. *103*, 4948 (1981)
28. Corey, E. J. et al.: ibid. *100*, 8034 (1978)
29. Oppolzer, W., Fröstl, W.: Helv. Chim. Acta *58*, 590 (1975)
30. Keck, G. E. et al.: Tetrahedron Lett. *22*, 2615 (1981)
31. Oppolzer, W. et al.: Helv. Chim. Acta *63*, 555 (1980)
32. Piers, E., Morton, H. E.: J. Org. Chem. *45*, 4263 (1980)
33. Pine, S. H. et al.: J. Am. Chem. Soc. *102*, 3270 (1980)
34. Pornet, J.: Tetrahedron Lett. *22*, 453 (1981)
35. Dressaire, G., Langlois, Y.: ibid. *21*, 67 (1980)
36. Kametani, T. et al.: J. Am. Chem. Soc. *103*, 1256 (1981)
37. Tsai, D. J. S., Matteson, D. S.: Tetrahedron Lett. *22*, 2751 (1981)
38. Syn Comm. *11*, 709
39. Jabri, N. et al.: Tetrahedron Lett. *22*, 959 (1981)
40. Grubbs, R. H. et al.: ibid. 2425 (1974)
41. Corey, E. J., Glass, R. S.: J. Am. Chem. Soc. *89*, 2600 (1967)
42. Brieger, G.: ibid. *85*, 3783 (1963)
43. Olsson, T., Wennerstöm, O.: Tetrahedron Lett. 1721 (1979)
44. McBee, E. T. et al.: J. Am. Chem. Soc. *84*, 4540 (1962)
45. Oppolzer, W., Snowden, R. L.: Tetrahedron Lett. 3505 (1978)
46. Frater, G.: Helv. Chim. Acta *57*, 172 (1974)
47. Näf, F. et al.: ibid. *60*, 1196 (1977)
48. Bichan, D. J., Yates, P.: Can. J. Chem. *53*, 2054 (1975)
49. Tavares, R. F., Katten, E.: Tetrahedron Lett. 1713 (1977)
50. Buchi, G. et al.: J. Org. Chem. *42*, 3323 (1977)
51. Cupas, C. A., Schumann, W.: J. Am. Chem. Soc. *92*, 3237 (1970)
52. Cupas, C. A. et al.: ibid. *93*, 4623 (1971)
53. Parker, K. A., Adamchuk, M. R.: Tetrahedron Lett. 1689 (1978)

54. Best, W. M., Wege, D.: ibid. *22*, 4877 (1981)
55. DeClerq, P. J., Van Royen, L. E.: Syn. Commun. *9*, 771 (1979)
56. Bilovic, D. et al.: Tetrahedron Lett. 2071 (1964)
57. Snowden, R. L.: ibid. *22*, 97 (1981)
58. Schiehser, G. A., White, J. D.: J. Org. Chem. *45*, 1864 (1980)
59. Ciganek, E.: ibid. *45*, 1512 (1980)
60. Greuter, H., Schmid, H.: Helv. Chim. Acta *57*, 1204 (1974)
61. Ciganek, E.: J. Am. Chem. Soc. *103*, 6261 (1981)
62. a) Oppolzer, W.: Synthesis 793 (1978)
    b) Kametani, T.: Pure Appl. Chem. *51*, 747 (1979)
    c) Funk, R. L., Vollhardt, K. P. C.: Chem. Soc. Rev. *9*, 41 (1980)
    d) Oppolzer, W.: Heterocycles *14*, 1615 (1980)
63. Ramamurthy, V., Liu, R. S. H.: J. Org. Chem. *39*, 3435 (1974)
64. Oppolzer, W.: J. Am. Chem. Soc. *93*, 3833 (1971)
65. Kametani, T. et al.: ibid. *98*, 3378 (1976)
66. Funk, R. L., Vollhardt, K. P. C.: ibid. *102*, 5253 (1980)
67. a) Nicolaou, K. C., Barnette, W. E.: J. C. S. Chem. Commun. 1119 (1979)
    b) Oppolzer, W. et al.: Helv. Chim. Acta. *62*, 2017 (1979)
68. Oppolzer, W., Keller, K.: Angew. Chem. Int. Ed. (Eng.) *11*, 728 (1972)
69. Quinkert, G. et al.: ibid. *19*, 1027 (1980)
70. Ito, Y. et al.: J. Am. Chem. Soc. *102*, 863 (1980)
71. Djuric, S. et al.: ibid. *102*, 6885 (1980)
72. Kaneko, C. et al.: Tetrahedron Lett. 1645 (1980)
73. a) Klemm, L. H. et al.: J. Org. Chem. *41*, 2571 (1976)
    b) Klemm, L. H., Gopinath, K. W.: Tetrahedron Lett. 1243 (1963)
74. Kuehne, M. E. et al.: J. Org. Chem. *45*, 3259 (1980)
75. Kotsuki, H. et al.: Chem. Lett. 917 (1981)
76. Klemm, L. H., Gopinath, K. W.: J. Het. Chem. *2*, 225 (1965)
77. Bartlett, A. J. et al.: J. C. S. Perkin I 1315 (1975)
78. Laird, T. et al.: ibid. 1477 (1980)
79. Kraus, G. A. et al.: Tetrahedron Lett. 853 (1979)
80. Houlihan, W. J. et al.: J. Org. Chem. *46*, 4515 (1981)
81. Frater, G.: Tetrahedron Lett. 4517 (1976)
82. Nakamura, Y. et al.: Helv. Chim. Acta *60*, 247 (1977)
83. Becker, H.-D. et al.: J. Org. Chem. *44*, 1336 (1979)
84. Martin, R. H. et al.: Helv. Chim. Acta *58*, 776 (1975)
85. Wasserman, H. H., Keehn, P. M.: Tetrahedron Lett. 3227 (1969)
86. Roush, W. R., Hall, S. E.: J. Am. Chem. Soc. *103*, 5200 (1981)
87. Roush, W. R., Peseckis, S. M.: ibid. *103*, 6696 (1981)
88. Gras, J.-L., Bertrand, M.: Tetrahedron Lett. 4549 (1979)
89. Bajorek, J. J. S., Sutherland, J. K.: J. C. S. Perkin I 1559 (1975)
90. Widmer, U. et al.: Helv. Chim. Acta *61*, 815 (1978)
91. Bazan, A. C. et al.: Tetrahedron *34*, 3005 (1978)
92. Kametani, T. et al.: J. C. S. Chem. Commun. 400 (1976)
93. Morgans, D. J., Stork, G.: Tetrahedron Lett. 1959 (1979)
94. Williams, D. J. et al.: Tetrahedron *36*, 3571 (1980)
95. Corey, E. J., Danheiser, R. L.: Tetrahedron Lett. 4477 (1973)
96. Roush, W. R., Gillis, H. R.: J. Org. Chem. *45*, 4283 (1980)
97. Zsindely, J., Schmid, H.: Helv. Chim. Acta. *51*, 1510 (1968)
98. Miyashi, T. et al.: Tetrahedron Lett. 155 (1979)
99. Shinmyozu, T. et al.: Chem. Lett. 405 (1978)
100. Mechoulam, R.: J. Am. Chem. Soc. *90*, 2418 (1968)

101. Chapman, O. L. et al.: ibid. *93*, 6696 (1971)
102. Oude-Alink, A. M. et al.: J. Org. Chem. *38*, 1993 (1973)
103. Snider, B. B. et al.: J. Am. Chem. Soc. *101*, 6023 (1979)
104. Tietze, L.-F., Kiedrowski, G. V.: Tetrahedron Lett. *22*, 219 (1981)
105. Snider, B. B., Duncia, J. V.: J. Org. Chem. *45*, 3461 (1980)
106. Tietze, L.-F. et al.: Angew. Chem. Int. Ed. (Eng.) *19*, 134 (1980)
107. Ciganek, E.: J. Org. Chem. *45*, 1497 (1980)
108. Hasan, I., Fowler, F. W.: J. Am. Chem. Soc. *100*, 6696 (1978)
109. Davies, L. B. et al.: J. C. S., Perkin I. 1293 (1978)
110. Davies, L. B. et al.: J. C. S. Chem. Commun. 663 (1977)
111. Jacobi, P. A., Craig, T.: J. Am. Chem. Soc. *100*, 7748 (1978)
112. Jojima, T. et al.: Chem. Pharm. Bull. *20*, 2191 (1972)
113. Garanti, L., Zecchi, G.: Tetrahedron Lett. *21*, 559 (1980)
114. Oppolzer, W.: Angew. Chem. Int. Ed. (Eng.) *11*, 1031 (1972)
115. Butsugan, Y. et al.: Tetrahedron Lett. 1129 (1971)
116. Wenkert, E., Naemura, K.: Syn. Commun. *3*, 45 (1973)
117. Roush, W. R., Gillis, H. R.: J. Org. Chem. *45*, 4267 (1980)
118. Bellville, D. J. et al.: J. Am. Chem. Soc. *103*, 718 (1981)
119. Ballivet-Tkatchenko, D. et al.: ibid. *101*, 2763 (1979)

Chapter 2.

The Intramolecular Diels-Alder Reaction: Reactivity and Stereocontrol

I. Introduction

Chapter One covers the range of dienes and dienophiles that have been used in the intramolecular Diels-Alder reaction, with some comments about the methods by which they can be prepared. This chapter addresses two other areas of interest: the factors that govern the rate of the intramolecular Diels-Alder reaction, and, in reactions leading to the formation of one or more new chiral centers, the factors that govern the stereochemical outcome of the cyclization.

II. Factors Influencing the Rate of Cyclization

There are four primary factors influencing the rate of a given intramolecular Diels-Alder reaction: dienophile substitution, diene substitution (including geometry), ring size (number of atoms bridging the diene and dienophile), and substituents on the bridge. These are considered in turn. It should be noted that, as for the intermolecular reaction, investigators are prone to report merely that a particular reaction proceeds, with little, if any, kinetic data. Comparison of one cyclization with another, therefore, will often be at best qualitative.

1. Influence of Dienophile Substitution on the Rate of Cyclization

The influence of dienophile substitution on the rate of the intermolecular Diels-Alder reaction has been investigated (1, 2). From this data, it can be concluded that an electron-withdrawing group will activate a dienophile in the order shown [1]. The same order of reactivity obtains for the intramolecular cyclization [2–4].

$$\text{[1]}$$

(3) [2]

(4) [3]

(5) (6) [4]

In addition to these monosubstituted examples, an excellent comparison of dienophiles bearing more than one electron-withdrawing group has appeared (4). Still, this area is not yet entirely understood (8). It should be pointed out that some trienes cyclize so rapidly ([5]-in this case, probably under acid catalysis) that kinetic studies would be difficult.

$$\text{(9) [5]}$$

It is apparent from the literature (1, 2) that α-activated dienophiles (meth-acryloyl) are more reactive in the intermolecular cyclization that β-activated dienophiles (crotyl). While there may be other factors involved, this seems also to be true for the intramolecular cyclization [6]. It is clear that increasing the alkyl substitution on the dienophile retards the cyclization [7].

$$\text{(3) [6]}$$

$$\text{(11) (10) [7]}$$

37

(12) [7]

2. Influence of Diene Substitution on the Rate of Cyclization

The accelerating effect of electron-donating diene substituents on the rate of the intermolecular Diels-Alder reaction is well known (1). It is apparent than an alkoxy substituent on the diene also accelerates the intramolecular cyclization [8]. As might be expected, this effect is not observed with a less polarized dienophile [9].

(13)(10) [8]

(14) [9]
(15)

Steric bulk encumbering the approach of the dienophile to the diene can substantially slow the cyclization [10]. Holding the diene cisoid by inclusion in a ring accelerates the reaction slightly [11]. Finally, there has been conflicting evidence about the reactivity of Z vs. E dienes. While it was initially observed (19) that Z and E dienes react at about the same rate, we (3) and others have found E dienes to be more reactive. Two recent studies [12, 13], taken together, show that, in fact, Z vs. E diene reactivity is a function of the particular geometry involved.

(16), (17) [10]

(18) [11]

(20) [12]

39

(21) [13]

3. Influence of the Elements Bridging the Diene and Dienophile on the Rate of Cyclization

In general, connecting the diene and the dienophile by a flexible bridge should make the reaction between them more facile. There are other factors, however, that may be as important, or more important, than this more favorable entropy: ring strain, buttressing, and conformational preferences, especially of hetero atoms.

a) Effect of Ring Size

For unactivated carbocyclic systems [14], 6/5 cyclization proceeds more readily than 6/6 cyclization. For activated systems in which the electron-withdrawing group is at the end of the dienophile away from the bridge [15], there seems to be little preference for 6/5 over 6/6. Finally, for activated systems in which the electron withdrawing group is at the end of the dienophile toward the bridge [16], or is included in the bridge [17], 6/6 cyclization is much more facile than 6/5 cyclization.

(22) [14]
(23)

MeO$_2$C

CO$_2$Me

150°
15 h

OSiMe$_3$

OSiMe$_3$

(24) [15]
(7)

CO$_2$Me

CO$_2$Me

150°
19 h

OSiMe$_3$

O SiMe$_3$

CH=O

HO

PhCH$_3$
reflux
I hour

O H

OH

H

(3) [16]
(25a)

CH=O

Ph

110°
18 h

O H O

Ph

O

190°
13 h

O

(11) [17]
(3)

O

RT
18 h

O

OTHP

OTHP

b) Effect of Buttressing

It would be expected that substituents on the bridge that tend to hold the diene and dienophile closer to each other would enhance the rate of the intramolecular Diels-Alder reaction. This effect has been demonstrated quantitatively by Boeckman [18], who found that a gem-dimethyl substitution on the bridge led to about a four-fold increase in the rate of the reaction (Thorpe-Ingold effect). This same effect is observed when the bridge is fused to a ring [19]. N-alkylation of an amide bridge, leading to an increased population of the amide conformer that can cyclize, has the same effect [20].

(4) [18]

(23) [19]

(26) [19]

(27) [20]

c) Effect of Heteroatoms in the Bridge on the Rate of Cyclization

There are two questions one might ask with regard to hetero-bridged systems. The first, the relative rate of cyclization from one ring size to another with a given bridge, has been addressed in part. For amine and ester bridged systems, 6/5 cyclization [21, 22] is faster than 6/6. For an amide bridged system [23, 24], there seems to be little difference. Although a few ethers have been cyclized (29a), no ring size preference data is available.

(28) [21]

(29a) [22]

from thermal conjugation of the diene

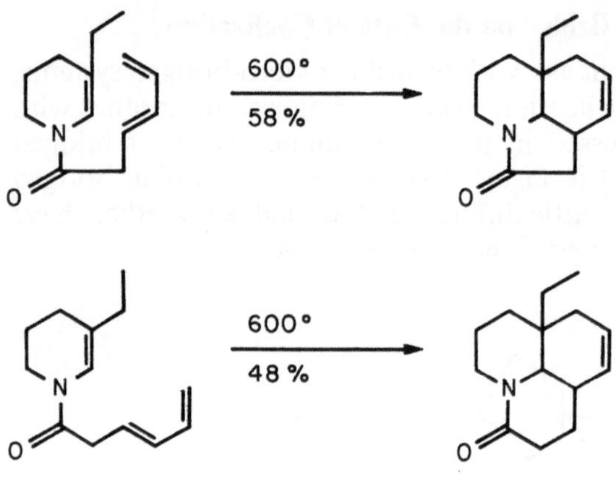

(30) [23]

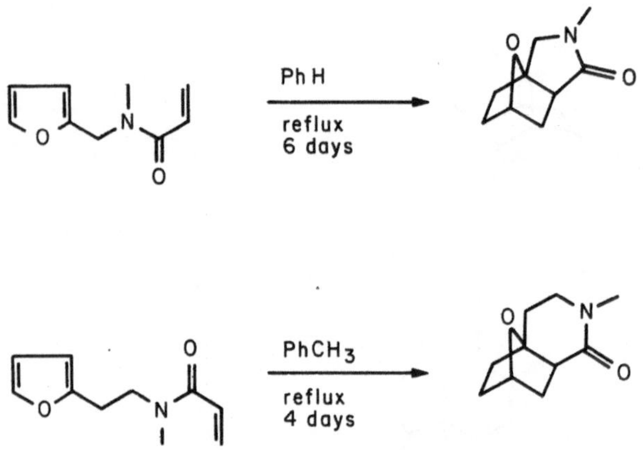

(6) [24]

The other pertinent information one might seek would be the effect on the rate of cyclization of including a heteroatom in the place of a methylene in the bridge. It is apparent that ethers are about the same as their all-carbon counterparts [25], while amides, with and without the carbonyl group in the bridge, are substantially slower [26]. Both this effect and the ring size effects noted above will probably vary with the position of the heteroatom in the bridge.

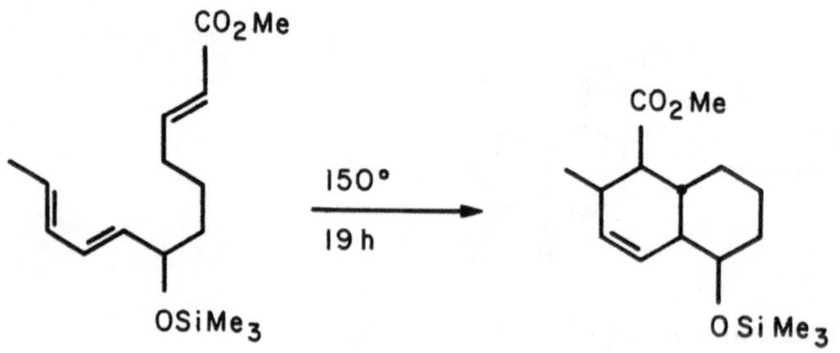

(7) [25]

44

CO₂Me

170°
18–22 h

MeO₂C

(29) [25]

160°

(23) [26]

190°
24 h

CO₂Me

CO₂Me

(31) [26]

190°

(31) [26]

III. Factors Influencing the Stereochemical Outcome of the Cyclization

When new asymmetric centers are created in the course of a cyclization, it is important to understand the factors that determine the configuration of those centers relative to each other, as well as to any asymmetric centers in the triene before cyclization. There are three elements that are important: triene geometry, cis vs. trans ring fusion, and diastereomeric control by remote chiral centers.

1. Triene Geometry

It is generally observed that olefin geometry is maintained in the course of the cyclization [27]. Exceptions have been noted when olefin isomerization at elevated reaction temperature preceded cyclization (29b).

(32) [27]

(27) [27]

2. Cis vs. Trans Ring Fusion

There are a number of factors influencing cis vs. trans ring fusion. These can be grouped into three general categories: preference for an endo transition state, influence of diene substitution, and influence of the bridge between the diene and the dienophile.

a) Preference for an Endo Transition State

The endo rule, formulated for the intermolecular Diels-Alder reaction, is not rigorously followed even in the intermolecular case (33). For intramolecular cyclizations, geometric factors (vide infra) tend to be more important. With highly activated dienophiles, and especially under acid catalysis, endo cyclization can predominate (34) [28, 29]. In the second example, the role of the acidic reaction medium in overcoming the geometric preference for an exo (chair-like) transition state is apparent. Other examples of improved endo selection on acid catalysis have been reported (7, 38).

(9) [28]

III. Factors Influencing the Stereochemical Outcome of the Cyclization

(3) [29]

oxidant		
BaMnO₄	43	57
MnO₂	66	34
PDC	84	16

55 : 45

> 95 : 5

(40) [30]

47

b) Influence of Diene Substitution on the Geometry of Ring Fusion

Wilson [30] has pointed out that free rotation around the single bond connecting the diene to the bridge can lead to a mixture of cis and trans products. A substituent at the 3-position of the diene can destabilize one of the rotamers, leading to improved selectivity in the cyclization (see p. 47).

c) Influence of the Bridge Between the Diene and the Dienophile on the Geometry of Ring Fusion

As the bridge between the diene and the dienophile is shortened from four to three atoms, the geometric outcome of the cyclization shifts, although the direction of the shift cannot yet generally be predicted [31–33]. It is apparent that inclusion of sp² centers in the bridge in conjugation with the diene or dienophile will increase the proportion of cis product [34, 35]. This outcome is predicted by molecular models if it is assumed that overlap between adjacent π orbitals is maintained.

III. Factors Influencing the Stereochemical Outcome of the Cyclization

(40) [32]

(22) [32]

(7) [33]

(7) [33]

$$55 : 45 \qquad (40)\ [34]$$

$$37 : 25 \qquad (31)\ [34]$$

$$\text{cis only} \qquad (31)\ [34]$$

$$(23)\ [35]$$

$$2 : 1 \qquad (41)\ [35]$$

3. Diastereomeric Control by Remote Chiral Centers

When the triene before cyclization contains one or more chiral centers, a new complexity is introduced. Rather than just two (cis vs. trans) possible products from the cyclization, there are now four. Through analysis of the transition states leading to cyclization, it is possible to design trienes that cyclize to give largely a single product.

a) Diastereomeric Control by a Rigid Skew in the Bridge

The efficacy of a rigid skew in the bridge between the diene and the dienophile in directing the diastereomeric outcome of the intramolecular Diels-Alder reaction was first demonstrated by Kametani [36]. It was previously known that cyclization via the o-quinone methide would proceed to give the B–C trans ring fusion (for reviews of o-quinone methide based intramolecular Diels-Alder reactions, see reference 43). The diastereomeric control in this cyclization derives from the 2,3 trans attachment of the diene and the dieno-phile to the cyclohexane ring. We have taken advantage of this same effect in a simple preparation of angularly substituted perhydrophenanthrenes [37]. Effective skew of the diene and the dienophile can be maintained even when one of the olefins of the diene is included in the ring. This has been demonstra-ted in the heterocyclic series by Corey [38] and in the carbocylic series by Stork [39].

(42) [36]

(26) [37]

(44) [38]

(38) [39]

b) Diastereomeric Control by a Substituent on the Bridge

There are early reports [40, 41] of intramolecular Diels-Alder reactions of diastereomeric mixtures, one component of which reacted more readily than the other. These can be rationalized by assuming that in the transition state leading to 6/6 cyclization, the bridging atoms will be staggered in their relationship one to another, and thus will assume a chair-like conformation.

One diastereomer would have the secondary methyl group equatorial on the chair, and would cyclize, to give the stereochemical outcome observed. The transition state leading to cyclization for the other diastereomer would be destabilized by the secondary methyl being axial. This principle has been applied to trienes having a single substituent on the bridge [42–44]. It is clear that the degree of chiral induction is related to the steric bulk of the substituent.

(45) [40]

(46) [41]

(23) [42]

(47) [43]

82 : 9 : 5 : 4

(stereochemistry of minor
diastereomers not determined)

(48) [44]

Substantial chiral induction can be seen even when the transition state leading to cyclization is not chair-like [45, 46]. The critical requirement is that there be substantial steric hindrance in one of the two alternative diastereomeric transition states leading to cyclization. This same analysis has been applied to heterobridged trienes [47, 48], and to the intramolecular cyclization of heterodienes [49]. Note, in either case, that the initial chiral center is inducing the chirality of the cyclized system. Given the ready availability in high optical purity of acyclic systems having a single chiral center, this opens a way to the enantioselective synthesis of polycyclic natural products (32a, 49).

(9) [45]

(32) [46]

(49) [47]

(21) [48]

(50) [49]

c) Diastereomeric Control by an Alkoxy Substituent on the Bridge

The lesser steric requirement of an alkoxy group makes it generally less effective than an alkyl group for directing the diastereomeric outcome of a cyclization [50, 51]. On the other hand, there are stereoelectronic effects with alkoxy substituents that can be used to advantage. Funk [52] has demonstrated, in a Lewis acid-catalyzed cyclization, a preference for an alkoxy group to be perpendicular to the dienophile in the transition state leading to cyclization, giving to rise a product in which the alkoxy group is axial. Franck (51) has reported a similar, thermal, effect in the intermolecular series. Note, however, that the results of Roush [53] in a very similar 6/5 cyclization are not predicted by this model.

(32) [50]

(37) [51]

| R = | Ph | 51 : 31 |
| R = | SiMe3 | 31 : 48 |

(39b) [52]

Via: or

55

(25b) [53]

d) Diastereomeric Control by Chiral Centers not Included in the Bridge

In the foregoing discussion, diastereomeric control is achieved by making one face of the diene or the dienophile sterically less accessible than the other. It follows that any substituent that effectively hinders only one face of the diene or the dienophile should exert diastereomeric control [54, 55]. Note that in the latter example, either face of the dienophile can be made the more hindered, depending on the choice of cyclization solvent.

(52) [54]

MeOH :	70 : 30
Benzene :	23 : 77

(53) [55]

If the asymmetric fragment exerting diastereomeric control were removable, and were itself optically pure, the result would be an enantioselective cyclization. Chiral acrylates have been used in the intermolecular Diels-Alder reaction, one of the more effective to date being phenylmenthyl (54). Roush has shown [56] that phenylmenthyl esters exert substantial diastereomeric control in the intramolecular Diels-Alder reaction also.

86 : 14

IV. Summary

The intramolecular Diels-Alder reaction has two particular strengths. On the one hand, thoughtful design of the starting triene can lead, specifically, to a highly functionalized product, as exemplified by the cytochalasin model studies of Fuchs [57]. On the other hand, simple linking of a diene to a dienophile followed by cyclization can be a quick way to assemble a desired skeleton [58]. These two strengths were combined early on in Chapman's elegant synthesis of carpanone [59].

(55) [59]

The treatment in this and the preceeding chapter are not meant to close the book on the intramolecular Diels-Alder reaction. Rather, the intent is to summarize and systematize early findings. As exemplified by two recent contributions [60, 61], there are yet many opportunities for creative application of such cyclizations.

(56) [60]

(57) [61]

V. References

1. a) Sauer, J. et al.: Chem. Ber. *97*, 3183 (1964)
   b) Blain, M. et al.: Tetrahedron *36*, 2775 (1980)
2. Sauer, J., Sustmann, R.: Angew. Chem. Int. Ed. (Eng) *19*, 779 (1980)
3. Taber, D. F. et al.: unpublished observations, Vanderbilt University
4. Boeckman, R. K., Jr., Ko, S. S.: J. Am. Chem. Soc. *104*, 1033 (1982)
5. DeClerq, P. J., Van Royen, L. E.: Syn. Commun. *9*, 771 (1979) ·
6. Parker, K. A., Adamchuk, M. R.: Tetrahedron Lett. 1689 (1978)
7. Roush, W. R., Hall, S. E.: J. Am. Chem. Soc. *103*, 5200 (1981)
8. White, J. D., Sheldon, B. G.: J. Org. Chem. *46*, 2273 (1981)
9. Taber, D. F., Gunn, B. P.: J. Am. Chem. Soc. *101*, 3992 (1979)
10. Ichihara, A. et al.: ibid. *102*, 6353 (1980)
11. Bajorek, J. J. S., Sutherland, J. K.: J. C. S. Perkin I, 1559 (1975)
12. Kraus, G. A., Taschner, M. J.: J. Am. Chem. Soc. *102*, 1974 (1980)
13. Schiehser, G. A., White, J. D.: J. Org. Chem. *45*, 1864 (1980)
14. Stork, G., Morgans, D. J., Jr.: J. Am. Chem. Soc. *101*, 7110 (1979)
15. Keck, G. E., Boden, E.: Tetrahedron Lett. *22*, 2615 (1981)
16. Tavares, R. F., Katten, E.: ibid. 1713 (1977)
17. Oppolzer, W., Snowden, R. L.: ibid. 3505 (1978)
18. Frater, G.: ibid. 4517 (1976)
19. House, H. O., Cronin, T. H.: J. Org. Chem. *30*, 1061 (1965)
20. Boeckman, R. K., Jr., Alessi, T. R.: J. Am. Chem. Soc. *104*, 3216 (1982)
21. Fuchs, P. L. et al.: ibid. *102*, 5960 (1980)
22. a) Jung, M. E., Halweg, K. M.: Tetrahedron Lett. *22*, 3929 (1981)
    b) Bal, S. A., Helquist, P.: ibid. *22*, 3933 (1981)
23. Wilson, S. R., Mao, D. T.: J. Am. Chem. Soc. *100*, 6289 (1978)
24. Roush, W. R.: J. Org. Chem. *44*, 4008 (1979)
25. a) Taber, D. F. et al.: Tetrahedron Lett. *22*, 5141 (1981)
    b) Roush, W. R., Peseckis, S. M.: J. Am. Chem. Soc. *103*, 6696 (1981)
26. Taber, D. F., Saleh, S. A.: ibid. *102*, 5085 (1980)
27. Gschwend, H. W. et al.: J. Org. Chem. *38*, 2169 (1973)
28. Greuter, H., Schmid, H.: Helv. Chim. Acta *57*, 1204 (1974)
29. a) Boeckman, R. K., Jr., Demko, D. M.: J. Org. Chem. *47*, 1789 (1982)
    b) Borch, R. F. et al.: J. Am. Chem. Soc. *99*, 1612 (1977)
30. Martin, S. F. et al.: ibid. *102*, 3294 (1980)
31. Oppolzer, W., Fröstl, W.: Helv. Chim. Acta *58*, 590 (1975)
32. a) Nicolaou, K. C. et al.: J. Am. Chem. Soc. *103*, 6967 (1981)
    b) Roush, W. R., Myers, A. G.: J. Org. Chem. *46*, 1511 (1981)
    c) Ley, S. V. et al.: Tetrahedron Lett. 361 (1981)
33. Berson, J. A. et al.: J. Am. Chem. Soc. *84*, 297 (1962)
34. It should be noted that if enolization toward the ring fusion is possible, subsequent epimerization can obscure the stereochemistry of the initial cyclization. Gras (35) has shown that MnO_2 in $CHCl_3$ is sufficient to effect such epimerization. Note also that a report by Oppolzer (36a) of exo cyclization has been withdrawn (36b).
35. Gras, J.-L., Bertrand, M.: Tetrahedron Lett. 4549 (1979)
36. a) Oppolzer, W., Snowden, R. L.: ibid. 4187 (1976)
    b) Oppolzer, W. et al.: Helv. Chim. Acta *64*, 2002 (1981)
37. Roush, W. R.: J. Am. Chem. Soc. *102*, 1390 (1980)
38. Stork, G. et al.: J. Am. Chem. Soc. *103*, 4948 (1981)

39. a) Roush, W. R. et al.: ibid. *104*, 2269 (1982)
    b) Funk, R. L., Zeller, W. E.: J. Org. Chem. *47*, 181 (1982)
40. Wilson, S. R., Huffman, J. C.: ibid. *45*, 560 (1980)
41. Wilson, S. R., Mao, D. T.: ibid. *44*, 3093 (1979)
42. Kametani, T. et al.: J. Am. Chem. Soc. *98*, 3378 (1976)
43. a) Oppolzer, W.: Synthesis 793 (1978)
    b) Kametani, T.: Pure Appl. Chem. *51*, 747 (1979)
    c) Funk, R. L., Vollhardt, K. P. C.: Chem. Soc. Rev. *9*, 41 (1980)
    d) Oppolzer, W.: Heterocycles *14*, 1615 (1980)
44. a) Corey, E. J., Danheiser, R. L.: Tetrahedron Lett. 4477 (1973)
    b) Corey, E. J. et al.: J. Am. Chem. Soc. *100*, 8034 (1978)
45. Fukamiya, N. et al.: J. C. S. Perkin I, 1843 (1973)
46. Naf, F., Ohloff, G.: Helv. Chim. Acta *57*, 1868 (1974)
47. Taber, D. F., Saleh, S. A.: Tetrahedron Lett. *23*, 2361 (1982)
48. Wilson, S. R. et al.: J. Org. Chem. *47*, 747 (1982)
49. Oppolzer, W., Fröstl, W.: Helv. Chim. Acta *58*, 593 (1975)
50. Tietze, L.-F., Kiedrowski, G. v.: Tetrahedron Lett., 219 (1981)
51. Franck, R. W. et al.: J. Am. Chem. Soc. *104*, 1106 (1982)
52. Mukaiyama, T., Iwasawa, N.: Chem. Lett., 29 (1981)
53. Kuehne, M. E. et al.: J. Org. Chem. *47*, 1335 (1982)
54. a) Ensley, H. E. et al.: ibid. *43*, 1610 (1978)
    b) Oppolzer, W. et al.: Tetrahedron Lett. *22*, 2545 (1981)
55. Chapman, O. L. et al.: ibid. *93*, 6696 (1971)
56. Deutsch, E. A., Snider, B. B.: J. Org. Chem. *47*, 2862 (1982)
57. Snider, B. B., Phillips, G. B.: J. Am. Chem. Soc. *104*, 1113 (1982)

Chapter 3.

The Intramolecular Alder Ene Reaction

I. Introduction

In comparison to the intramolecular Diels-Alder reaction, the intramolecular Alder ene reaction has been little studied (1, 2). All examples in the literature through mid-1981 are included in the Tables at the end of this chapter.

For the purpose of this chapter, the ene reaction is considered to be the transfer of a proton from an ene donor to an ene acceptor, with concomitant formation of a bond between them. In the intramolecular case, R and R' will be joined [1]. The aim of this chapter is to present the scope and limitations of the intramolecular ene reaction as they now stand, and to suggest directions to be explored that might make this reaction even more useful than it has been in organic synthesis.

$$\text{Donor} \quad \text{Acceptor} \quad \xrightarrow{\text{ENE}} \quad \text{Product} \qquad [1]$$

Nature of the Reaction

Formally, the ene reaction is a concerted, six-electron process. In practice, as with some examples (see Chapter Two) of the intramolecular Diels-Alder reaction, nonsynchronous bond formation may play an important role [2]. With some exceptions (3), free radical inhibitors have no effect on the reaction (4).

$$[2]$$

The Intramolecular Alder Ene Reaction

Electron withdrawing substituents on the ene acceptor accelerate the reaction, and this effect is accentuated by Lewis acid catalysis. The corollary, that electron-donating substituents on the ene donor could also accelerate the reaction, has not been explored.

Orientation of the Reaction

Depending on the positioning of the bridge linking the ene donor and acceptor, three orientations are possible for the intramolecular ene reaction. These have been described by Oppolzer (1) as Types I, II, and III. Snider (5) has observed, in addition, Type IV [3]. Type I reactions are by far the most studied, especially for five- and six-membered ring forming reactions. It is conceivable, especially as activated ene systems are developed, that Types II–IV will become more important.

[3]

Type I   Type II   Type III   Type IV

Bridge Length

An intramolecular ene reaction that leads to formation of a five-membered ring is almost always more facile than the analogous reaction leading to a six-membered ring. Seven-membered ring formation is generally not an efficient process. This is in contrast to the intramolecular Diels-Alder reaction, where there is not always such a clear-cut preference for five over six. This difference can be rationalized by considering the transition states for the two reactions. In the case of the Diels-Alder reaction, torsional strain in the shorter bridge can sometimes make it difficult to achieve the requisite orbital overlap between the diene and the dienophile. In the ene reaction, only one carbon-carbon bond is formed. Orbital alignment need not be as rigidly maintained, and the geometrically more accessible five-membered ring is favored.

62

II. Reactivity of Ene Acceptors

Ene acceptors can be divided into three classes: olefinic, unactivated; olefinic, activated by conjugation with electron withdrawing groups; and acetylenic. Examples of each class are listed in the Tables.

1. Unactivated Olefins

This is the least reactive but most studied class of ene acceptor. When the other functionality in the molecule is thermally stable, cyclizations with such acceptors can very clean [4, 5]. In the third example [6], the rigid polycyclic framework holding the donor and acceptor parallel and overlapping facilitates the reaction. Problems may arise when the prospective acceptor has an allylic proton and can function as a donor. Two products are then possible, and the ratio between them will depend on the relative reactivity of the competing donor/acceptor combinations [7]. Note that [4–6] are Type I reactions. Type III and Type IV reactions would also be possible in [4] and [5], but are not observed. In [7], the competition is between two different Type III pathways.

(6) [4]

(7) [5]

(8) [6]

(9) [7]

63

2. Activated Olefins

The most common electron-withdrawing group used for acceptor activation has been alkoxycarbonyl. As would be expected, addition of Lewis acid further activates the acceptor [8]. At elevated temperatures, ene adducts can be equilibrated [9]. In this case, equilibration may be taking place by enolization, rather than by retro-ene followed by recyclization. Aldehydes and ketones can also be used to activate acceptors [10]. In the case shown, further activation with Lewis acids gave largely non-ene products.

(10) [8]

$$- OR -$$

$$\frac{3 \text{ of } Et_2 AlCl}{CH_2Cl_2 \quad -35°/30\,min}$$

(11)

(1) [9]

$$(12) \ [10]$$

There is a competing possibility with carbonyl activated acceptors, that the acceptor can function as a Diels-Alder diene. In practice, this problem has been overcome by raising the reaction temperature to a point at which the Diels-Alder reaction is reversible [11].

$$(13) \ [11]$$

3. Acetylenes

Acetylenes appear to be more reactive acceptors than olefins [12], perhaps because orbital overlap is easier to achieve with the cylindrically symmetrical acetylene. Cyclization conditions are mild enough that highly functionalized systems can be tolerated [13] (15). As with olefins, electron-withdrawing groups further activate acetylenic acceptors [14] (14).

$$(14) \ [12]$$

$$(15) \ [13]$$

65

$$95\% \qquad (14)\ [14]$$

III. Reactivity of Ene Donors

1. Unactivated Donors

The ene donors used to date have almost exclusively been unactivated olefins. When such olefins have more than one proton available to donate, mixtures of products may arise. As illustrated, it would appear that the reaction shows some preference for formation of the E olefin [13], but little discrimination between the transfer of a methyl vs. a methylene hydrogen [15].

$$(15)\ [13]$$

$$(13)\ [15]$$

2. Speculations on Activated Donors

As electron-withdrawing groups activate ene acceptors, it seems plausible that electron-donating groups could activate ene donors. This aspect of the intramolecular ene reaction has not been systematically investigated. The

66

most extreme case would be the intramolecular addition of an allyl anion to an olefin. Such cyclizations have been studied, and have been found to be more facile than the cyclization of the corresponding saturated secondary anion, even though the latter is less stable and should be more reactive [16]. Other examples are yet to be explored. For instance, an appealing case can be made that the ene reaction should show the same alkoxide activation as the Cope (17) and vinyl cyclopropane (18) rearrangements [17]. Oppolzer [18] recently reported several synthetic applications of the anionic ene reaction.

(16) [16]

[17]

(19) [18]

IV. Steric and Stereoelectronic Control Elements in the Intramolecular Ene Reaction

1. Cis vs. Trans Relationship of the Two Interacting Side Chains

In general, five-membered ring formation proceeds to give the two side chains cis [19], while six-membered ring formation proceeds to give the side chains trans [20]. This is easily rationalized by the need for the two side chains to remain close one to another as the cyclization progresses.

(4, 20) [19]

57 : 19

(21) [20]

2. Diastereometric Control by Substituents on the Bridge Between the Ene Donor and the Ene Acceptor

The diastereomeric control illustrated above [19, 20] can also easily be rationalized, by considering nonbonding interactions in the alternative transition states leading to cyclization. In the six-membered case [21], the transition state leading to cyclization would be chair-like. That transition state in which the pendant alkyl group was equatorial on the incipient chair would be favored. In the five-membered case [22], the interactions are more subtle, and diastereomeric control may not be as clean. It should be noted that with an acetylenic acceptor, it would seem necessary, on consideration of models [23], to employ a Z-donor to get clean diastereomeric control.

[21]

[22]

[23]

3. Diastereomeric Control of Carbon-Carbon Bond Formation by Other Remote Substituents

As would be expected, a chiral ester that effectively hinders one face of the ene acceptor will substantially induce the chirality of the ring as it forms [24]. The Lewis acid-catalyzed reaction may be proceeding through a transition state in which the catalyst is complexed both to the enone and to the π-donating benzene ring. The role of the catalyst in the reaction is two-fold: held between the arene and the carbonyl, [25], it effectively blocks one face of the unsaturated ester while, at the same time, activating the ester as an ene acceptor.

$$\begin{array}{c} \text{Et}_2\text{AlCl} \\ \xrightarrow{\hspace{2cm}} \\ -35° \\ 60\% \end{array}$$

95 : 5

$$\xrightarrow{\hspace{1cm}\Delta\hspace{1cm}}$$

1 : 1

$$\begin{array}{c} \text{Et}_2\text{AlCl} \\ \xrightarrow{\hspace{2cm}} \\ -35° \end{array}$$

15 : 85

(22) [24]

[25]

Enantiomer

V. Directions for the Future

The possibility of developing activated ene donors is discussed above. As active donors are coupled with active acceptors, it may be possible to develop routes to heretofore relatively inaccessible ring systems and substitution patterns. In particular, one or more of the interacting carbon atoms could be replaced with a heteroatom. Several such systems, as exemplified by the acyl nitroso acceptors of Keck [26], the Lewis acid-mediated carbonyl acceptors developed by Snider [27], and the enol donor cyclizations of Conia [28], have already been investigated.

(23) [26]

(24) [27]

(25) [28]

In addition to Type I reactions, which have received the most attention so far, the other cyclization modes may also become important. For instance, activated Type III reactions might offer a viable entry to substituted medium ring systems [29].

[29]

The Intramolecular Alder Ene Reaction

As activated donor-acceptor combinations are developed, allowing milder reaction conditions, it should be possible to cyclize more highly functionalized dienes and enynes. Analysis of the results of such cyclizations should lead to an understanding of the influence of different substituents on the stereochemical outcome, and thus, to the rational design of donor-acceptor combinations that would be expected to lead to a particular desired product. The intramolecular Alder ene reaction, leading as it does from simple acyclic functionality to a highly functionalized ring, should then be increasingly useful in the synthesis of complex natural products.

VI. Tables

Table 1. Carbocyclic, Type I, Unactivated Olefin Acceptor

Starting Material	Conditions	Product(s)	Yield	Ref.
	457° 56 sec		—	26
	—		—	26
	490°		75	27
	350°	 57% 19%	—	4, 27

Table 1 (continued)

Starting Material	Conditions	Product(s)	Yield	Ref.
	650° 15 torr	S.m. 37% 8.7% 4.8% 28% 15%	—	28
	290° 2.5 h	S.m. 20% 2% 40% 17% 5%	—	29
	450° 12 torr	S.m. 15% 9% 41% 25%	—	29
	350°	(CH=O products) 4:1	—	5

Table 1 (continued)

Starting Material	Conditions	Product(s)	Yield	Ref.
	500°		93	7
	180° 30 h		60	30
	345° 40 min		71	7
	290° 72 h	1.7:1	68	31

75

The Intramolecular Alder Ene Reaction

Table 1 (continued)

Starting Material	Conditions	Product(s)	Yield	Ref.
	$T_{1/2} = 6\,h$ RT		—	8
	380° 1 sec		80	1
	280° 24 h		17	32
			—	33

Table 2. Carbocyclic, Type I, Activated Olefin Acceptor

Starting Material	Conditions	Product(s)	Yield	Ref.
	300° 14 h		65	1
	300° 14 h		65	1
	570°	 6:1	70	34
	200°		90	1

77

78

Table 2 (continued)

Starting Material	Conditions	Product(s)	Yield	Ref.
	400°	CO₂Me three isomers	87	27
	RT		29	21
	240° 84 h	53:47	87	12
	MeAlCl₂ CH₂Cl₂ RT, 20 h		85	12

Table 3. Carbocyclic, Type I, Acetylene Acceptor

Starting Material	Conditions	Product(s)	Yield	Ref.
	400°		64	35
	225° 4 h		95	14
	200°		quantitative	36
	90° 20 h		quantitative	14

79

The Intramolecular Alder Ene Reaction

Table 3 (continued)

Starting Material	Conditions	Product(s)	Yield	Ref.
	225° 48 h		17	14
	225° 48 h		15	14
	135° 24 h		95	14
	225° 62 h		85	14

Table 3 (continued)

Starting Material	Conditions	Product(s)	Yield	Ref.
	260° 2.5 h		80	37
	250° 27 min	 7:2 E/Z	60	15

81

The Intramolecular Alder Ene Reaction

Table 4. Heterocyclic, Type I, Unactivated Olefin Acceptor

Starting Material	Conditions	Product(s)	Yield	Ref.
	280° 7 h		88	6
	280° 7 h	89:11	—	6
	210° 2 h		43	6
	290° 27 h		—	6

Table 4 (continued)

Starting Material	Conditions	Product(s)	Yield	Ref.
	280° 5 h		84	6
	270° 7 h	1:1	—	6
	230° 63 h	99.8:0.2	82	6
	230° 63 h	74:26	90	6
	230° 1 min		93	38

The Intramolecular Alder Ene Reaction

Table 4 (continued)

Starting Material	Conditions	Product(s)	Yield	Ref.
	290° 13 h		75	6
	250° 3 h		80	6
	230° 42 h		55	6
	230° 42 h	84.5:15.5	55	6

Table 5. Heterocyclic, Type I, Activated Olefin Acceptor

Starting Material	Conditions	Product(s)	Yield	Ref.
SiMe$_3$–N, O, MeO$_2$C (butenoyl dienamide)	150° 24 h	SiMe$_3$–N pyrrolidinone, CO$_2$Me, isopropenyl	—	1
SiMe$_3$–N, O, CO$_2$Me	300° 16 h	SiMe$_3$–N pyrrolidinone, CO$_2$Me, isopropenyl	60	1
CF$_3$CO–N, CO$_2$Et	220° 30 min	CF$_3$CO–N pyrrolidine, CO$_2$Et, isopropenyl (two diastereomers) 86:14	81	39

85

The Intramolecular Alder Ene Reaction

Table 5 (continued)

Starting Material	Conditions	Product(s)	Yield	Ref.
	135° 210 h		30	13
	135° 210 h		43	13
	120° 30 h	1:1	95	13
	AlCl$_3$ CH$_2$Cl$_2$ RT 1 h	1:1	29	40

Table 5 (continued)

Starting Material	Product(s)	Conditions	Yield	Ref.
(CF₃ acyl, N, CO₂Me, CO₂Me enamide structure)	(CF₃ acyl pyrrolidine, CO₂Me, CO₂Me, isopropenyl structure)	180° 35 h	90	41
(N-isopropyl, OEt fumarate amide, cyclohexene structure)	(isopropyl lactam with CH₂CO₂Et, fused cyclohexene structure)	150°, distill	79	38
(CF₃ acyl, E, CO₂Et, prenyl structure)	(CF₃ acyl pyrrolidine, E, E, CO₂Et, isopropenyl structure)	80° 24 h	97	10
(CF₃ acyl, E, E, CO₂Et, prenyl structure)	(F₃C acyl pyrrolidine, E, E, CO₂Et, isopropenyl structure)	3 eq Et₂AlCl CH₂Cl₂ −35°/30 min	90	11
(F₃C acyl, E, E, CO₂Et, prenyl structure)	(F₃C acyl pyrrolidine, E, E, CO₂Et, isopropenyl structure) and (F₃C acyl pyrrolidine, E, E, CO₂Et, isopropenyl structure) 89:11	3 eq Et₂AlCl CH₂Cl₂ −35°/6 h	78	11

The Intramolecular Alder Ene Reaction

Table 5 (continued)

Starting Material	Conditions	Product(s)	Yield	Ref.
	Δ	1:1		22
	Et_2AlCl $-35°$	95:5	60	22
	Et_2AlCl $-35°$	15:85	—	22

Table 5 (continued)

Starting Material	Conditions	Product(s)	Yield	Ref.
(structure with CO$_2$Me, CO$_2$Me)	95—110° 40 h	(lactone with CO$_2$Me, MeO$_2$C)	45	13
(bicyclic structure with CO$_2$Me, CO$_2$Me)	RT	(structure with CO$_2$Me, CO$_2$Me)	—	42
(structure with CO$_2$Me, CO$_2$Me)	125° 6 h	(structure with CO$_2$Me, CO$_2$Me) + (lactone with CO$_2$Me, CO$_2$Me) 3:2	66	13
(structure with CO$_2$Me, CO$_2$Me)	125° 6 h	2:3	70	13

The Intramolecular Alder Ene Reaction

Table 6. Heterocyclic, Type I, Acetylene Acceptor

Starting Material	Conditions	Product(s)	Yield	Ref.
	180° 5 h		50	1

Table 7. Type II

Starting Material	Conditions	Product(s)	Yield	Ref.
	280° 7 h		40	1
	210° 2 h		95	43
	245° 6 h		78	43
	260° 5 h		60	43
	225° 3 h		quant	43
	212° 6 h		quant	43

Table 8. Type III

Starting Material	Conditions	Product(s)	Yield	Ref.
	330° 13 h		90	3
	290° 76 h	+ isomer 9:1	84	44
	300° 90 h		80	44
	350° 24 h	+ 1:1	quant	9
	270—320°	+ 4:1	—	9

Table 9. Organometallic

Starting Material	Conditions	Product(s)	Yield	Ref.
(structure with MgBr)	$T_{1/2} = 2$ h RT	(structure with MgBr)	—	16
(structure)	Buli TMEDA	(structure with Li)	30	45
(structure with Cl)	1) Mg powder 2) 60°, 23 h 3) acrolein	(structure with OH)	57	19a
(structure with Cl)	1) Mg powder 2) RT, 20 h 3) O_2	(structure with OH)	70	19a
(structure with Cl)	1) Mg*, −78° 2) 50°, 16 h 3) $CO_2/80°$, 2 h	(structure with CO_2H)	47	19b

VII. References

1. Oppolzer, W., Snieckus, V.: Angew. Chem. Int. Ed. (Eng.) *17*, 476 (1978)
2. Hoffman, H. M. R.: ibid. *8*, 556 (1969)
3. Lambert, J. B., Napoli, J. J.: J. Am. Chem. Soc. *95*, 294 (1973)
4. Huntsman, W. D., Curry, T. H.: ibid. *80*, 2252 (1958)
5. Snider, B. B., Duncia, J. V.: J. Org. Chem. *45*, 3461 (1980)
6. Oppolzer, W., Pfenninger, E., Keller, K.: Helv. Chim. Acta *56*, 1807 (1973)
7. Plavac, F., Heathcock, C.: Tetrahedron Lett. 2115 (1979)
8. Brown, J. M.: J. C. S. (B) 868 (1969)
9. Lambert, J. B., Fabricius, D. M., Napoli, J. J.: J. Am. Chem. Soc. *101*, 1793 (1979)
10. Oppolzer, W., Andres, H.: Tetrahedron Lett., 3397 (1978)
11. Oppolzer, W., Robbiani, C.: Helv. Chim. Acta *63*, 2010 (1980)
12. Snider, B. B., Rodini, D. J., van Straten, J.: J. Am. Chem. Soc. *102*, 5872 (1980)
13. Snider, B. B., Roush, D. M., Killinger, T. A.: ibid. *101*, 6023 (1979)
14. Snider, B. B.: J. Org. Chem. *43*, 2161 (1978)
15. Stork, G., Kraus, G. A.: J. Am. Chem. Soc. *98*, 6747 (1976)
16. Felkin, H., Umpleby, D., Hagaman, E., Wenkert, E.: Tetrahedron Lett., 2285 (1972)
17. Evans, D. A., Golob, A. M.: J. Am. Chem. Soc. *97*, 4765 (1975)
18. Danheiser, R. L., Martinez-Davila, C., Morin, J. M., Jr.: J. Org. Chem. *45*, 1340 (1980)
19a. Oppolzer, W., Battig, K.: Tetrahedron Lett. *23*, 4669 (1982)
    b. Oppolzer, W. et al.: ibid. *23*, 4673 (1982)
20. Pines, H., Hoffman, N. E., Ipatieff, V. N.: J. Am. Chem. Soc. *76*, 4412 (1954)
21. Tietze, L.-F., Kiedrowski, G. V.: Tetrahedron Lett. *22*, 219 (1981)
22. Oppolzer, W., Robbiani, C., Battig, K.: Helv. Chim. Acta *63*, 2015 (1980)
23. Keck, G. E., Webb, R.: Tetrahedron Lett. 1185 (1979)
24. Karras, M., Snider, B. B.: J. Am. Chem. Soc. *102*, 7951 (1980)
25. Bortolussi, M., Bloch, R., Conia, J. M.: Bull. Soc. Chim. Fr. 2722 (1975)
26. Huntsman, W. D., Solomon, V. C., Eros, D.: J. Am. Chem. Soc. *80*, 5455 (1958)
27. Huntsman, W. D., Lang, P. C., Madison, N. L.: J. Org. Chem. *27*, 1983 (1962)
28. Strickler, H., Ohloff, G.: Helv. Chim. Acta *50*, 759 (1967)
29. Schulte-Elte, K. H., Gadola, M., Ohloff, G.: ibid. *54*, 1813 (1971)
30. Mauer, W., Grimme, W.: Tetrahedron Lett. 1835 (1976)
31a. Oppolzer, W.: Helv. Chim. Acta *56*, 1812 (1973)
    b. Oppolzer, W., Mahalanabis, K. K.: Tetrahedron Lett. 3411 (1975)
    c. Oppolzer, W., Mahalanabis, K. K., Battig, K.: Helv. Chim. Acta *60*, 2388 (1977)
32. Oppolzer, W., Battig, K., Hudlicky, T.: ibid. *62*, 1493 (1979)
33. Padwa, A., Rieker, W. F.: J. Am. Chem. Soc. *103*, 1859 (1981)
34. Mayer, C. F., Crandall, J. K.: J. Org. Chem. *35*, 2688 (1970)
35. Huntsman, W. D., Hall, R. P.: ibid. *27*, 1988 (1962)
36. Naegeli, P., Kaiser, R.: Tetrahedron Lett. 2013 (1972)
37. Townsend, C. A., Scholl, T., Arigoni, D.: J. Chem. Soc. Chem. Commun. 921 (1975)
38. Schmitz, E. et al.: J. Prakt. Chemie *321*, 387 (1979)
39. Kennewell, P. D., Matharu, S. S., Taylor, J. B., Sammes, P. G.: J. C. S. Perkin I, 2542 (1980)
40. Snider, B. B., Roush, D. M.: J. Org. Chem. *44*, 4229 (1979)

VII. References

41. Oppolzer, W., Andres, H.: Helv. Chim. Acta *62*, 2282 (1979)
42. Kelly, T. R.: Tetrahedron Lett. 437 (1973)
43. Wender, P. A., Letendre, L. J.: J. Org. Chem. *45*, 367 (1980)
44. Marvell, E. N., Cheng, J. C.-P.: ibid. *45*, 4511 (1980)
45. Edwards, J. H., McQuillen, F. J.: J. Chem. Soc. Chem. Commun. 838 (1977)

Subject Index

Reactivity and Structure

Concepts in Organic Chemistry

Editors: K. Hafner, J.-M. Lehn, C. W. Rees,
P. v. R. Schleyer, B. M. Trost, R. Zahradník

Springer-Verlag
Berlin
Heidelberg
New York
Tokyo

Lecture Notes in Chemistry

Editors: G. Berthier,
M. J. S. Dewar, H. Fischer,
K. Fukui, G. G. Hall,
H. Hartmann, H. H. Jaffé,
J. Jortner, W. Kutzelnigg,
K. Ruedenberg, E. Scrocco

Springer-Verlag
Berlin
Heidelberg
New York
Tokyo

Volume 34: **N. D. Epiotis**
Unified Valence Bond Theory
of Electronic Structure – Applications
1983. VIII, 585 pages. ISBN 3-540-12000-9

Volume 33: **G. A. Martynov, R. R. Salem**
Electrical Double Layer at a Metal-dilute
Electrolyte Solution Interface
1983. VI, 170 pages. ISBN 3-540-11995-7

Volume 32: **H. F. Franzen**
Second-Order Phase Transitions and the
Irreducible Representation of Space Groups
1982. VI, 98 pages. ISBN 3-540-11958-2

Volume 31: **H. Hartmann, K.-P. Wanczek**
Ion Cyclotron Resonance Spectrometry II
1982. XV, 538 pages. ISBN 3-540-11957-4

Volume 30: **R. D. Harcourt**
Qualitative Valence-Bond Descriptions of
Electron-Rich Molecules:
Pauling "3-Electron Bonds" and
"Increased-Valence" Theory
1982. X, 260 pages. ISBN 3-540-11555-2

Volume 29: **N. D. Epiotis**
Unified Valence Bond Theory of Electronic
Structure
With collaboration of J. R. Larson, H. L. Eaton
1982. VIII, 305 pages. ISBN 3-540-11491-2

Volume 28: **G. S. Ezra**
Symmetry Properties of Molecules
1982. VIII, 202 pages. ISBN 3-540-11184-0

Volume 27: **W. Bruns, I. Motoc, K. F. O'Driscoll**
Monte Carlo Applications in Polymer Science
1981. V, 180 pages. ISBN 3-540-11165-4